Bringing Back the Beast:
Restoration Strategies for Sasquatch Habitats

RAY DAVID

&

JEREMY LATCHAW

Copyright © 2023 BRINGING BACK THE BEAST
All rights reserved

DEDICATION

This book is dedicated to our families and friends who have endured our actual searches for Sasquatch as well as the countless hours of research making this book possible. You all know who you are but we must at least mention; Molly, Ruthann, Danny, Xander, Debbie, Shelly, Candi, Cheri, Holly and Ramon II.

CONTENTS

Preface

Chapter 1: Who is Sasquatch   1

Chapter 2: The Sasquatch Habitat   8

Chapter 3: Endangered Species in North America   14

Chapter 4: Our Experience Looking for Sasquatch   21

Chapter 5: Environmental Reconstruction   27

Chapter 6: Monitoring Changes in the Environment   32

Chapter 7: Strategies for Restoring Habitats   37

Chapter 8: Potential Challenges   41

Chapter 9: Benefits of Restoration Efforts   46

Chapter 10: Sasquatch Conservation Efforts   51

References   56

Acknowledgements   57

About the Authors   58

PREFACE

Sasquatch's History

The reporting of large creatures that walk upright goes back several hundred years. For a shorter period of time, about 60-70 years, reports of human-like foot imprints have been reported. Some of these imprints have been photographed and even casts of the imprints have been made. It should also be noted that sightings are not limited to years in the past. Sightings continue to be made; some by casual campers and weekend hikers as well as by professionals who make a living from being out in the environment.

Many Native American cultures have oral histories that tell of a primate-type creature roaming the continent's forests. In these tales, the animals are sometimes more human-like and, other times, more ape-like. The oral history seems to have begun with the Kwakiutl indigenous peoples in what is now southwestern Canada and northwestern United States. They called her "Dzunukwa" and described Sasquatch as a big, hairy female that lives deep in the forests. Their history says that she was rarely seen because she was caring for her children and was always in a protection mode rather than a confrontation mode. When it came to a decision of "fight or flight", she would rather slip away than be seen. She was also referred to with another name, "Sasquatch". That name was used by the Halkomelems, who were members of the central branch of the Salishan family of languages, in southern British Columbia and northern Washington State. There are three dialect groups there, located from west to east, Island, Downriver and Upriver (SavingSasquatch.org, 2022).

Further south, on the Tule River Indian Reservation near Porterville, California, stands a collection of large boulders in an area known as 'Painted Rock'.

Alcove opening at Painted Rock

In a stone alcove on the grounds, one can observe a collection of pictographs portraying the 'Hairy Man', as well as two other beings of similar appearance. The largest depiction is that of a hulking figure standing on two legs, covered in moderately long fur, with intense eyes which appear to be 'dripping', likely an artistic rendition of crying. The large figure is accompanied by two others, one slightly smaller than the first and the other much smaller (Strain, 2012).

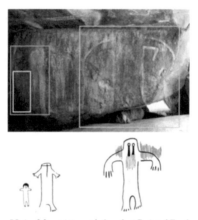

Hairy Man pictograph found at Painted Rock

It would be conceivable that these smaller figures could represent 'Hairy Woman' and 'Hairy Child', if the naming convention were carried out that far. These pictographs are estimated to be several hundred years old, and are in very poor condition due to weathering and vandalism.

Additional details can be found at www.SavingSasquatch.org

1

WHO IS SASQUATCH

Sasquatch, or more commonly known as Bigfoot, has long been a part of folklore and mystery. For centuries, reports of sightings of large, hairy bipedal humanoids have emerged from wilderness areas around the world. The creature is often described as being between 6-8 feet tall with big eyes, long arms, and an ape-like face. It is largely believed to be a species of undiscovered hominids, though this remains unverified by science. Whatever the truth behind it may be, Sasquatch has become an iconic figure in popular culture and continues to fascinate those who have heard or experienced its presence.

Overview of Sasquatch – origin, description

In Native American folklore, Sasquatch is often depicted as a powerful and mysterious creature. According to legend, the Sasquatch was once believed to be an ancient spirit of the forest that protected its lands from intruders. Stories are told of how it could make itself invisible or transform into other animals in order to avoid detection by humans. In some stories, the Sasquatch even helped lost travelers find their way out of the woods by leaving behind a trail of giant footprints for them to follow. These mythical tales have been passed down through generations as a reminder that nature still holds many mysteries yet to be discovered.

Experts believe that the Sasquatch is a reclusive species of hominid thought to have evolved from an ape-like ancestor. It is considered to be very shy, and rarely seen in the wild. It has been reported to move quickly and silently through forests, making it difficult for people to spot them. While there are no confirmed sightings of these creatures by zoologists, many experts believe they do exist because of the numerous reports from reliable sources over the years. There is still much mystery surrounding this creature and its existence remains a source of debate among scientists and cryptozoologists alike.

The debate on the existence of Sasquatch continues to this day, with a split opinion among experts. While some remain skeptical, many believe that there is enough evidence to suggest the possibility of its existence. Reports from reliable sources have led some experts to estimate that around 60% of cryptozoologists and 30% of scientists believe in the existence of Sasquatch. This number is likely higher as more people become aware of these reports and stories. Despite this, without hard proof or an actual sighting, it will be difficult for either side to prove their case one way or the other.

Sasquatch sightings have been reported in many locations around the world, although they are often concentrated in remote and heavily wooded areas. Reports of Sasquatch sightings indicate that they can be seen most frequently during twilight hours and at night. Eyewitnesses often describe the creature as being incredibly swift and agile, suggesting that it is well adapted to its habitat. Sightings have also suggested that Sasquatch has the ability to vocalize with a variety of noises such as growls, whistles, and screams.

The Sasquatch population is believed to be decreasing due to habitat destruction from human encroachment on its natural environment, as well as hunting and illegal poaching for fur or other body parts. While there is still debate about its legitimacy as an actual species, it is widely accepted among researchers that the Sasquatch should be protected because of its potential value to science if further evidence is found to support its existence.

History of Sasquatch Sightings

In 1924, a Canadian newspaper published an article featuring an account of a large ape-like creature with red eyes observed near Harrison Hot Springs in British Columbia. Since then, there have been numerous accounts of Sasquatch sightings throughout North America, as well as other parts of the world like China, Russia, and even Papua New Guinea.

The earliest recorded Sasquatch sighting is said to have occurred in the 1820s near Mount St. Helens in Washington State, USA. According to reports, a group of Native American hunters encountered the creature and described it as being over seven feet tall and covered with reddish-brown hair. It had a strong, distinct odor and was reported to be walking on two legs like a human being. Witnesses also noted that it seemed to be able to vanish into the forest quickly when startled. While this sighting did not gain much attention at the time, it is now considered one of the most important historical accounts of Sasquatch activity and a crucial piece of evidence for modern-day believers.

Patterson film, Bluff Creek, California, 1967

One of the most talked about Sasquatch sightings in history occurred in 1967 near Bluff Creek, California. A construction worker by the name of Roger Patterson filmed a creature that many believe to be a Sasquatch walking across an open field and into nearby woods. This event has been widely discussed over the years as it is one of the few pieces of evidence that suggests Sasquatch may actually exist. Scientists have studied this film for decades and have yet to come up with a definitive answer regarding its authenticity. However, believers continue to cite it as proof that these creatures are real, citing its clear footage and lack of any known hoaxes or special effects at the time.

The most recent Sasquatch group sighting in the United States occurred in North Carolina. A group of hikers reported seeing a large, black creature with yellow eyes and covered in short, shaggy fur. It had a distinctly human-like face but was very large in size and walked on two legs. The hikers described the creature as standing over seven feet tall and being incredibly strong and agile. They were unable to capture any evidence of their encounter due to the creature's swiftness, but they did manage to snap a few photographs from a distance. After further investigation into the incident, experts determined that it was likely an undocumented species related to Sasquatch. This sighting is one of several that have been reported throughout the country in recent years, adding to the mystery surrounding these elusive creatures.

Where Sasquatch Sightings Occur

Sasquatch sightings are reported in various parts of the world every year. According to one report, in the United States alone there were over 3,000 sightings between 1913 and 2021. Canada also reports a significant number of Sasquatch sightings annually with an average of 200 reports per year. In addition, China has seen a notable increase in Sasquatch sightings since the early 2000s, resulting in roughly 500 reported sightings each year. Reports from Russia and Papua New Guinea make up the remainder with approximately 100 reported sightings annually.

Sasquatch sightings are most commonly reported in remote, rural areas with dense vegetation or forested terrain. They have also been reported near rivers and lakes, as well as mountainous regions. Reports of Sasquatch sightings typically occur during the night or twilight hours when visibility is low, making it easier for the creature to remain hidden from people's sight.

The United States has seen a considerable amount of Sasquatch sightings over the years, with certain regions having more reported sightings than others. The Pacific Northwest region is arguably one of the most popular hotspots for Sasquatch sightings in the country, with Washington and Oregon having some of the highest number of reports. California also sees a significant number of Sasquatch sightings each year, although not as many as its neighboring states to the north. Other notable hotspots include Colorado, Wyoming, and Montana where numerous encounters have been reported over time.

The Great Lakes region of North America is no exception when it comes to Sasquatch sightings. Reports of the creature have been coming from this area since the late 1800s, and more recently in the 2000s. In addition, Sasquatch have also been spotted near rivers and mountainous regions within this region.

A map showing United States Sasquatch sightings.

"Sasquatch" sightings have also been reported in the southern United States, particularly in states such as Arkansas, Louisiana and Mississippi. It should be noted that while there are reports of Sasquatch activity in this region, they are much lower than other parts of North America due to their more populated nature.

Why Sasquatch is Endangered

The exact number of Sasquatch creatures in the world is unknown, and it is believed that their population is declining due to loss of habitat and human encroachment. The creature's behavior often leads to conflict with people, resulting in increased hunting pressure. Additionally, the increasing number of Sasquatch sightings may be causing them to become more elusive as they strive to avoid human contact. As a result, Sasquatch is considered an endangered species by some experts who believe its numbers are dwindling.

In the last decade, human encroachment on wetlands has had a serious and detrimental impact on the environment. Wetlands (a preferred Sasquatch habitat) are essential for biodiversity conservation and play an important role in water purification, flood control, groundwater recharge, erosion prevention and carbon sequestration. Despite their importance to both humans and wildlife alike, wetlands have been disappearing at an alarming rate due to increased urbanization, industrial activities and agricultural practices. As a result of this rapid destruction of wetland ecosystems, numerous species are being pushed closer to extinction while others struggle just to survive.

Wetlands are essential to Sasquatch, as they provide the creature with a safe haven to hide and roam. Wetlands offer a diverse array of habitats which Sasquatch needs in order to sustain themselves. These habitats include marshes, swamps, bogs and fens, which are all important for their survival. Wetlands allow them access to food and water sources, such as fish and small mammals that live in the wetlands, as well as providing shelter from predators. Additionally, wetlands also offer protection against extreme temperatures by providing a layer of insulation between the cold air above and the warm ground below. Furthermore, wetlands can also provide camouflage for Sasquatch given its dense foliage and tall grasses. The importance of wetlands for Sasquatch cannot be understated — without them, it is likely that their populations would suffer greatly or even become extinct eventually.

Saving Sasquatch

Several organizations have worked toward officially listing Sasquatch as an endangered species in order to protect this unique creature from extinction. However, due to lack of scientific evidence and knowledge about the Sasquatch, they have not been recognized as such by government agencies. As a result, it is up to individuals and organizations to protect these creatures from harm by raising awareness and working toward their conservation.

In order to save this endangered species, it is essential to work toward bringing back the ecosystems they need in order to survive. Wetlands and dense forests are not only important for providing food and shelter, but also offer protection from extreme temperatures. Therefore, protecting these habitats is key to preserving the future of the Sasquatch as well as other wildlife that rely on wetlands and forests for survival.

If the wetlands and habitats that Sasquatch needs to survive are successfully restored, not only would this species be able to thrive again but it could also bring about scientific proof of its existence. This evidence would be crucial in supporting the case for Sasquatch's placement on the endangered species list. The restoration of wetlands would also offer so much more than just a habitat for Sasquatch – these ecosystems provide valuable benefits to humans and animals. For instance, wetlands are essential for biodiversity conservation, acting as filters to purify water, preventing floods and erosion, recharging groundwater and sequestering carbon emissions.

Ultimately, preservation of wetland and forest habitats would create an environment that not only supports Sasquatch but many other wildlife species as well. This could potentially lead to stronger scientific evidence of the creature's existence which may then be used to declare it an endangered species – thereby granting them protection from human harm and allowing them to thrive once again.

2

THE SASQUATCH HABITAT

While there are many theories about where Sasquatch live, the most likely habitats for them include dense forest areas, wetlands and mountainous regions. These environments provide ample food sources and shelter for these creatures to thrive in. Unfortunately, some of the species native to these areas may be endangered due to population encroachment on their habitats. In order to figure out how to discover Sasquatch, it will be helpful to discuss and explore the possible Sasquatch habitats and discuss what creatures in those environments are proven to exist.

Overview of Sasquatch Habitats

Sasquatch are believed to inhabit some of the most dense and remote forested areas, wetlands, and mountainous regions in North America. The natural habitats of these creatures are typically characterized by deep green forests, lush wetlands, and vast mountain ranges.

These areas often provide a habitat for a range of species including large carnivores (e.g., wolves, grizzly bears, cougars), deer, elk, moose, and a variety of smaller animals (e.g., raccoons and coyotes). In addition to these species that are thriving in these habitats, there are also some species that are endangered or threatened due to habitat destruction caused by population encroachment. These include species such as the spotted owl, northern goshawk, and wolverine.

In more recent years, human population encroachment on these habitats has increased significantly. This has resulted in a decrease in natural resources available to Sasquatch and other wildlife inhabitants of these areas. It also poses a threat to their survival as they are often forced into smaller habitats or even out of their native range altogether.

By examining the habitat requirements of Sasquatch, insights can be gained into what type of environment is ideal for them to live. From this, one can understand how to better protect these habitats and ensure that Sasquatch have a future in North America. Additionally, this knowledge can help inform conservation efforts to ensure the health of all wildlife species inhabiting these areas.

Dense Forest Areas

The United States northwest dense forests are home to some of the most diverse and unique habitats in the world. From the lush temperate rainforest of Olympic National Park in Washington, to the vast boreal forests and alpine tundra of Alaska, these forests provide an important habitat for many species including Sasquatch. Within these dense forests, many species thrive including old-growth conifers such as western hemlocks, Douglas fir, cedar and Sitka spruce. These trees provide shelter for various wildlife, including mammals such as black bear, deer and elk; birds like bald eagles and woodpeckers; amphibians such as salamanders; reptiles such as snakes; and fish like salmon and trout.

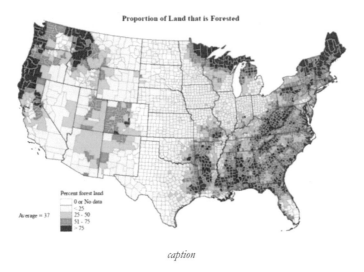

caption

Some interesting plants that can be found in these lush forests include deer ferns which have intricate fronds that resemble antlers; skunk cabbage which emits a pungent odor when it blooms; and Oregon grape with its bright yellow flowers that produce tart edible berries. The unique environment of these dense forests makes them a great place to explore and discover fascinating flora.

The Great Lakes region is home to some of the most stunning and diverse forests in the world. Spanning a total area of 600,000 square miles and containing more than 3,500 tree species, this forested region is remarkable for its vast array of habitats and ecosystems. From towering red oaks to ancient white pines and spruce, these forests provide essential habitat for numerous species of birds, mammals, reptiles and amphibians.

In addition to providing habitat for wildlife, the forests of the Great Lakes region are also important sources of timber and other resources used by humans. The timber industry is one of the largest employers in the region, providing thousands of jobs to local communities. In fact, over 2 billion board feet of timber are harvested each year from Great Lakes forests. These sustainable management practices allow for long-term economic prosperity while still preserving healthy forest ecosystems.

Furthermore, Great Lakes forests play an important role in sustaining healthy watersheds throughout the region. Trees cleanse pollutants from waterways by absorbing nutrients through their roots and filtering sediment from surface runoff with their leaves. Healthy forests also help store carbon dioxide in their trunks, where it can be safely stored away from the atmosphere. This helps counteract climate change effects such as increasing temperatures or sea level rise that would otherwise affect nearby communities.

Wetlands

North America is home to some of the most diverse and important wetlands in the world. Wetlands are ecosystems that exist between land and water, characterized by saturated soils, shallow standing water, or flowing water. In North America, wetlands occur in all regions of the continent and are found in coastal waters such as estuaries, marshes, bogs, swamps and shallow ponds.

Wetlands provide many benefits to people and wildlife. They act as buffers to storms and floods by absorbing excess water from heavy rainfall or melting snow. They also filter pollutants from runoff before it enters waterways. Wetlands play an important role in recharging groundwater supplies by replenishing aquifers with new fresh water each year. In addition, these ecosystems are a natural habitat for plants and animals; they serve as breeding grounds for many species of fish and migratory birds use them along their migratory routes. Below is the map showing all the major locations for watersheds including their drainage and watershed area.

North American watershed and drainage

In North America alone there are more than two hundred types of wetlands. These range from freshwater marshes to river deltas to peat bogs located in Arctic tundra regions. Despite their varied locations across the continent, all these habitats share common characteristics including low nutrient levels due to slow decomposition rates, high plant diversity because of adaptive strategies like broad root systems or floating leaves, and unique microbial communities that help shape wetland ecology.

Mountainous Areas

North America is home to some of the most stunning and majestic mountain ranges in the world. From the Rocky Mountains of the United States and Canada to the Sierra Madre Occidental of Mexico, these mountains are a vital part of the continent's natural landscape.

The Rocky Mountains, stretching from British Columbia to New Mexico, form an impressive 3,000-mile long range with peaks ranging in elevation from 8,000 to over 14,000 feet. The Rockies are made up of several distinct subranges including the Sawatch Range, Mogollon Rim and San Juan Mountains. These ranges offer some spectacular scenery, as well as opportunities for outdoor recreation such as hiking, camping & skiing.

One of North America's longest mountain ranges is the Sierra Madre Occidental. Spanning more than 1,400 miles in western Mexico, this range reaches heights up to 8891 feet tall in its highest peak, Cerro Mohinora. This area is home to unique flora and fauna species such as cougars, wild turkeys and jaguars which can be seen roaming amongst the forests and canyons formed by this range. Additionally, majestic waterfalls can be found throughout this area with Basaseachi Falls being one of the most spectacular examples at almost 800 ft deep.

Species Thriving in These Areas

North American forests, wetlands, and mountainous areas are home to a variety of wildlife species that have adapted over time to these unique environments. In the Rocky Mountains of Canada and the United States, animals such as grizzly bears, moose, elk, cougars, wild turkeys, and jaguars can be found inhabiting the region. These species are well-suited for life in high-altitude environments and have long been a part of the landscape.

In addition to these creatures, wetland ecosystems provide a habitat for various species including beavers, ducks, geese, muskrats and otters. These animals are able to utilize their specialized adaptations to survive in marshy areas with access to plenty of food sources. Meanwhile forested areas attract a huge range of woodland creatures such as deer, foxes, squirrels, rabbits, skunks, hawks, and woodpeckers, to name a few.

The Sierra Madre Occidental is also home to several unique species that are well adapted for life in its steep terrain. Here one will find birds such as hummingbirds and quetzals as well as large cats like pumas and jaguars. This mountain range is also host to numerous smaller mammals such as coatis, kinkajous, raccoons and opossums. Many of these animals rely on the numerous caves throughout this area for shelter from predators or harsh weather conditions.

Population Encroachment on these Areas

The population encroachment of North American forests, wetlands, and mountainous areas is a growing concern due to the rapid development of human settlements and industry. In the Rocky Mountains, cities such as Denver, Colorado Springs, and Salt Lake City have grown significantly over the years, displacing wildlife from their natural habitats. In addition to this urban growth, mining and logging operations have resulted in significant habitat loss for the numerous species native to the region. As a result, many of these animals are now listed as endangered or threatened species.

Wetlands throughout North America are facing similar pressures from human development with pollution from industrial activities posing a major threat to animal populations. In addition to human development, deforestation through wildfires destroys habitats each year. The destruction of wetland ecosystems has led to drastic declines in species such as piping plovers and bald eagles which rely on these areas for food and shelter.

Conclusions about Sasquatch Habitats

It is clear that North American forests, wetlands and mountainous areas are home to a variety of species, withsome endangered or threatened due to population encroachment. In order to protect these habitats and the animals living in them, it is important for humans to be conscious of their impact on the environment. To ensure that Sasquatch sightings to remain possible now and into the future, preserving their native habitat must become a priority. Only with a collective effort it be possible that these mysterious creatures will thrive in their natural habitats.

Above all, it is important to remember that Sasquatch sightings are a reminder of the importance of preserving wilderness areas and animal habitats for future generations. It is crucial to continue conservation efforts so that this incredible species can remain part of the world for years to come.

Endangered Species in North America

Endangered species in North America are facing a variety of threats, from habitat destruction and degradation, to human activity that can have devastating effects on local ecosystems. By examining these factors, one can gain a better understanding of why certain animals (and Sasquatch) are becoming increasingly endangered and what steps need to be taken to protect them.

Current Endangered Species in Sasquatch Habitats

When discussing endangered species in Sasquatch habitats, it is important to consider the various environments these species inhabit. Wetlands, mountainous terrains and forests are all potential homes for a variety of different species, some of which may be endangered.

Wetlands are areas of land that are permanently or seasonally covered with water, which can include marshes, swamps and bogs. Wetlands provide essential habitat for many species including amphibians, fish, birds, reptiles and mammals. The presence of these species in wetlands is threatened due to human activities such as draining, filling, and pollution, which can damage or destroy the habitat.

Mountainous terrains offer a diversity of animal habitats due to their different elevations, slopes and vegetation zones. Species living in these areas tend to be more adapted to extreme conditions such as cold temperatures, high winds and low moisture levels. Forests are home to a wide variety of wildlife species including birds, mammals and reptiles.

The United States of America and its neighbor Canada are home to numerous species of animals and plants, including some which are considered endangered or threatened. There have been recent reports of Sasquatch sightings in the wilds of North America, and as this region is full of wildlife which need to be protected, here is a look at the top thirteen most endangered species in North America:

1. The Florida Panther is the official state animal of Florida, but sadly this beautiful big cat is considered one of the rarest mammals on earth with only 120 to 230 individuals still remaining in the wild. These felines face threats like habitat destruction and

fragmentation, as well as collisions with vehicles, poaching, and inbreeding.
2. The Hawaiian Monk Seal can only be found in the waters of Hawaii, but due to human activity like overfishing and pollution, the population has been decreasing for years with roughly 1,200 individuals still alive.
3. The Mexican Wolf is one of North America's smallest wolves, but it is also one of the most endangered. It was once abundant in parts of North America, but due to hunting, trapping, and poisoning, the population has decreased drastically with only around 200 individuals remaining in their natural habitats.
4. The California Condor is an impressive vulture that can have a wingspan up to 9 ft, but due to the effects of DDT and hunting, only around 500 individuals remain in North America. To help protect this species, they are now bred in captivity and released into the wild.
5. The Black-Footed Ferret is another animal that was once abundant throughout North America, but today it is considered one of the rarest mammals in the world with just around 300 individuals living in the wild. Habitat loss, poisoning, and disease are the biggest threats to their survival.
6. The Red Wolf is a critically endangered carnivore native to the southeastern United States, and it is one of the most endangered species in North America. Once abundant throughout the region, their population has decreased drastically due to habitat loss and hybridization with coyotes. Today, there are only about 200 individuals remaining in their natural habitats.
7. The Whooping Crane is the tallest bird in North America and one of the rarest birds on earth, with just around 750 individuals still alive in the wild. They face numerous threats such as habitat destruction, largely from human activity.
8. The American Pika is a small mammal found in high mountain regions of western North America, and is currently listed as a threatened species. Habitat destruction due to climate change, overgrazing by livestock, and increasing temperatures are the biggest threats.
9. The Vancouver Island Marmot is one of the most endangered mammals in North America, with only 30 individuals remaining in their natural habitats in Canada. Unfortunately their population continues to decline due to disease, reduced food availability

caused by logging, and predation by introduced species.
10. The Canada Lynx is a threatened wildcat native to North America, with only around 5000 individuals remaining in the United States and Canada. They face numerous threats to their survival including habitat loss due to logging and human development, trapping, illegal hunting, and disease.
11. The American Burying Beetle is a critically endangered species of beetle found in North America, with less than 1500 individuals left in the wild. They are threatened by habitat loss due to land development, as well as pesticide use which affects their food sources.
12. The Pallid Sturgeon is an ancient species of fish that has inhabited the Missouri and Mississippi rivers for centuries, but it is now listed as an endangered species due to overfishing, dams and other human-created obstacles. Currently there are only about 200 individuals left in the wild.
13. The Mountain Plover is a small migratory shorebird found in the western United States and Canada, with only around 2000 breeding individuals remaining across the continent. They face numerous threats including habitat loss due to agricultural development, climate change, and predation by various introduced species.

Assessing existing threats to endangered wildlife

Endangered wildlife are found in all parts of the world, but particularly in North America. Here, species such as Sasquatch and other rare animals face a number of threats from human activity and habitat destruction that could lead to their extinction. These things can include habitat loss, hybridization, human hunting/trapping, and disease.

Habitat loss is one of the main threats to endangered species across the United States. As humans continue to expand their land use activities and urbanize or develop previously wild areas, an increasing number of species are losing their natural habitats and becoming endangered.

Humans are rapidly expanding their land activities and urbanizing in North America. Urbanization has been increasing at a rapid pace since the mid to late 20th century, with the total population of major metropolitan areas in the United States alone, growing from under 200 million people in 1950 to nearly 330 million by 2019. The driving forces behind this growth include improvements in transportation infrastructure, access to resources, and the availability of jobs.

Urbanization has had a large impact on the environment in North America, with areas around cities becoming increasingly developed with new housing, businesses, and industries. This has caused increased air pollution from vehicles and factories as well as additional runoff from paved surfaces that can pollute waterways. Despite these negative impacts, urbanization in North America is likely to continue to grow at a rapid pace due to its many benefits. Urban areas offer access to high-paying jobs, access to cultural amenities such as museums and theaters, and access to a wide range of services such as healthcare and education.

Illegal hunting and trapping in America has been a serious issue for the last decade. According to the U.S. Fish and Wildlife Service, over the past ten years there have been an estimated 2,000 to 3,000 cases of illegal hunting and trapping each year. The most common forms of illegal hunting activities are hunting without a license, hunting out of season, and taking game in areas closed to hunting. These activities can have devastating impacts on wildlife populations as well as local ecosystems.

In addition to making it difficult for fish and wildlife populations to recover or remain stable, illegal hunting and trapping often disrupts natural habitats and can cause long-term damage to the environment. Illegal activities also put a strain on the legal hunting community and can lead to overharvesting of certain species, which can further deplete populations.

The U.S. government has taken several measures to help prevent illegal hunting activities over the last ten years. These include requiring hunters to obtain licenses for specific species in each state, increasing penalties for violations, and implementing stricter enforcement measures. Despite these efforts, illegal hunting activities continue to be a major issue. Conservation organizations such as the National Wildlife Federation have also taken steps to help protect wildlife by advocating for stronger laws and regulations, providing public education on responsible hunting practices, and working with state agencies to develop strategies to reduce poaching.

Illegal hunting of endangered species has been an increasingly serious problem over the last decade. According to a report from the U.S. Fish and Wildlife Service, 78% of reported violations in 2019 were related to illegal hunting of threatened or endangered species, with certain species more at risk than others. A majority of these cases involved migratory birds, with more than half of the violations involving bald eagles and other raptors. This can lead to a permanent population decline, as these species are already at risk of extinction and may not be able to recover from any additional mortality.

Disease can also be a major threat to endangered species as it can cause rapid population declines if unchecked. In recent years, some bat species have been affected by a disease known as the white-nose syndrome, which is caused by the fungus Pseudogymnoascus destructans and has decimated populations of bats across the continent. This fungal infection causes bats to wake up during hibernation, resulting in dehydration and starvation. The fungus has killed millions of bats since it was first identified in 2006 and is still spreading.

Other diseases such as the amphibian Chytrid fungus have caused devastating declines in many species of frogs and other amphibians. This disease can cause widespread population decline within a few weeks or months, and has been linked to the extinction of many species around the world.

In addition, reintroduction programs for critically endangered species like the black-footed ferret have encountered challenges due to diseases that can spread quickly among wildlife populations. For example, an outbreak of Sylvatic plague killed more than 80% of a reintroduced population of ferrets in South Dakota in 2017. As a result, the species remains one of the most endangered mammals in North America.

Sources of habitat destruction and degradation

Deforestation is one of the leading causes of habitat destruction, as it involves the clearing of large amounts of land for industrial use, such as infrastructure and agricultural production. This leads to drastic changes in the environment which can have serious consequences for habitats, including the loss of shelter and food sources for many species.

The proper way to deforest in order to protect the environment is to use a method known as selective logging, which involves harvesting only certain trees within a forest, ensuring that the ecosystem can remain largely intact and natural. This type of logging allows for a more sustainable type of deforestation, as it prevents the entire forest from being cleared away and provides habitats for existing species to remain intact and healthy. Additionally, selective logging helps to prevent soil erosion, which can otherwise lead to significant habitat degradation. By utilizing this method of deforesting, the environment can be protected from damage while still allowing for a certain degree of land use.

Pollution is another major cause of habitat destruction, as it can poison and contaminate natural resources, while also disrupting the food chain, and changing the chemistry of water systems. An example of pollution that caused destruction in a habitat is the Deepwater Horizon oil spill in the Gulf of Mexico in 2010. This disaster was caused by an explosion on an offshore drilling rig, which released more than 200 million gallons of crude oil into the gulf. This oil spill affected thousands of species across hundreds of miles, and damaged habitats both on land and in the sea. It caused the deaths of dolphins, fish, turtles, birds and other species, as well as a loss of food sources for many animals due to contamination. As a result of this disaster, it took years for the Gulf Coast habitats to recover from the destruction that was done.

Invasive species are non-native creatures that are introduced into new environments, often with devastating effects on native species. These invasives can compete with native animals and plants for resources, leading to their displacement or extinction.

One invasive species causing significant problems to a local ecosystem is the zebra mussel in North America's Great Lakes. These small, freshwater mussels were first seen in Lake St. Clair near Detroit, Michigan in 1988 and have since spread throughout the Great Lakes region. They reproduce quickly and can attach themselves to boats, docks, and other structures, allowing them to be easily transported to new bodies of water.

The zebra mussel population has had a major impact on the Great Lakes ecosystem. They consume large amounts of plankton which are essential for fish populations, and clog up intake pipes used by municipal water plants. This has caused damage to both the environment and the economy, as fishermen have seen a decrease in their catches and cities have had to invest heavily in new pipe systems. The zebra mussel population has also caused native mussel species to decline or disappear entirely.

As discussed earlier, urban sprawl, construction, and commerce all have major impacts on habitats, as they fragment or destroy them outright. These activities also often reduce the amount of natural resources available, making it difficult for species to survive.

The environment of North America is home to many endangered species, including Sasquatch. Unfortunately, there are many threats to these habitats, such as deforestation, pollution, and invasive species. To protect these endangered species it is essential that humans take action to reduce their impact on the environment by participating in sustainable practices and investing in restoration efforts. Additionally, it is important to be aware of invasive species and take steps to prevent their spread so that native species can have a chance at survival.

4

Our Experience Looking for Sasquatch

In the year 2019, a peculiar sasquatch sighting arose in Northern Michigan. Our acquaintance, Eric Farrand, fervently believing in the existence of these elusive creatures, reached out to Jeremy, a person with a professional background in drones and expertise in teaching first responders on locating lost individuals in heavily forested areas. The latter's unique skill set appeared to be useful in potentially tracing the creature's whereabouts. This was prompted by receipt of a report from high school students claiming to have spotted a sasquatch in the region. It is noteworthy to mention that this approach of utilizing airborne technology had not been implemented in previous instances of similar sightings. Jeremy notified Ray David, whose background in environmental science and physics would help lead the team from a scientific perspective.

The quest for Sasquatch in northern Michigan was not an endeavor taken lightly by our team. We assembled a diverse group of experts to aid in our search. To further enhance our team's capabilities, we recruited Seth Getz, a renowned tracker with a talent for interpreting subtle signs of animal presence. Through his vast experience and intuition, Seth was able to guide us along the path less traveled, leading us to follow a series of clues that ultimately brought us closer to our elusive target.

Our team was determined to document every aspect of our quest for Sasquatch in northern Michigan, making use of state-of-the-art cameras to capture even the slightest of details. This became the pilot episode for our series, "Saving Sasquatch," which aimed to shed light on the mystery surrounding these elusive creatures and provide valuable scientific insights into their behavior and habitat. We worked tirelessly to ensure that our documentation was of the highest quality, hoping to contribute to the ongoing dialogue in the world of zoology and environmental science.

When it comes to searching for Sasquatch, most teams are made up of dedicated enthusiasts with a strong interest in cryptozoology and a passion for adventure. These individuals believe in the existence of

Sasquatch and are committed to finding evidence to support their belief.

Many Sasquatch hunting teams use a systematic approach to their search, covering large areas of land and focusing on locations where sightings or other evidence of Sasquatch activity has been reported. Some teams also conduct interviews with locals and collect stories and legends that may offer clues about the creature's habits and behaviors. In stark contrast to the typical approach taken by Sasquatch hunting teams, our team adopted a more scientific methodology, leveraging advanced remote sensing techniques rather than relying on traditional hunting tactics.

It is estimated that millions of people in North America believe in Sasquatch's existence, with a significant number of them actively searching for evidence to support their assertions. A 2017 survey conducted by Chapman University in California found that 14% of Americans believed in the existence of Bigfoot. The same survey also reported that a higher number of people, around 21%, believe in ghosts, while only 7% believe in aliens.

Although many skeptics dismiss Sasquatch sightings as misidentification, hoaxes, or even hallucinations, the fact that thousands of sightings have been reported indicates that the belief in Sasquatch's existence is not uncommon. The search for evidence of Sasquatch continues to attract enthusiasts from various backgrounds, including scientists, amateur researchers, and adventurers.

Our approach consisted of analyzing satellite imagery, and thermal imaging, along with studying historical data to identify potential hotspots for Sasquatch activity. Once we had identified our target locations, we deployed drones with high-resolution cameras to scour the terrain for any hints of the elusive creature.

We were also meticulous in our search to collect and analyze any physical evidence found in the field, including hair samples, tracks, and fecal matter, to help paint a more complete picture of Sasquatch and its behavior. The initial concepts were to set up tracking techniques like other sasquatch searchers would do, including trail cameras, look for watering holes, food sources etc.

The Drones Used

Drones have become an increasingly popular tool for remote sensing in recent years, and our Sasquatch hunting team took advantage of this technology to aid in our search. By using drones equipped with high-resolution cameras, we were able to capture aerial footage of the areas we were investigating, providing us with a bird's-eye view of the terrain and the potential hotspots for Sasquatch activity.

One of the most significant benefits of using drones is the ability to cover a large area quickly and efficiently. Instead of conducting ground-level searches, which can be slow and time-consuming, drones can cover a significant distance in a short amount of time, allowing for a more thorough investigation of the landscape.

Drones are also highly versatile and can be equipped with a range of sensors to capture different types of data. For our Sasquatch hunting expedition, we used drones equipped with high-resolution cameras to capture detailed imagery of the landscape. Some drones can also be equipped with thermal imaging sensors, which can detect heat signatures, making them ideal for identifying animals or other creatures that may be hiding in the dense forest. (We used a DJI M200 with a FLIR XT2 thermal camera.)

Another significant advantage of drones is the level of detail they can capture in their images. Using high-resolution cameras, we were able to capture detailed images of the terrain, including any tracks or other physical evidence that may have been missed by ground-level searches. Our image resolution, depending on our altitude above the surface, averaged about one inch. On one drone image captured, one can see where there is a potential sasquatch on the ground, but we could not confirm identity. Drones can also be used to capture images from different angles, which can provide a more comprehensive view of the landscape and any potential Sasquatch activity.

Our First Drone Launch

The first drone launch was interrupted with drone control signal problems. The drone went out of control, glitching the monitors, and causing Jeremy to land it immediately. We could not figure out what exactly caused this dynamic. Normally there are few tech issues because there is typically very little interference from technology or other sources of electromagnetic radiation in such areas. In fact, remote areas can be some of the best places for remote sensing. This is because the lack of interference can lead to more clear and accurate data, particularly when it comes to detecting heat signatures or other subtle signs of animal activity. Furthermore, the fact that these areas are so difficult to access means that there is often very little human activity, which can further reduce the potential for electromagnetic interference. This makes remote areas ideal for Sasquatch observation, as it allows researchers to thoroughly investigate the landscape without worrying about external factors affecting their results.

First we believe that there could have been sprite lightning in the area. This phenomenon, also known as 'sprites', occurs in the atmosphere about 38-50 miles above the earth's surface, and is often associated with thunderstorms. This becomes interesting from a Sasquatch perspective because there are some physicists who have posed the theory that there may be an interdimensional functionality to these sprites due to their high electrical intensity. If one has the open mindset to consider the possibility of Sasquatch existence, it should come as no surprise that a select few would have no problem considering the potential for interdimensional mobility. The electrical discharge generated by sprites high above the surface produces a red-orange flash that lasts only a few milliseconds. Despite being rare and difficult to observe, sprites have been studied extensively by NASA, due to the potential interference with radio communications and the earth's atmosphere.

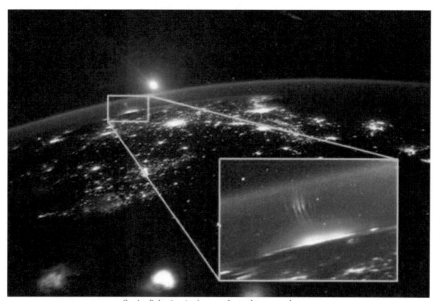

Sprite lightning jetting out from the atmosphere

One way in which sprite lightning can interfere with the spectrum is by generating a powerful electromagnetic pulse (EMP). This pulse can affect radio frequencies and create electromagnetic noise that can disrupt radio transmissions. Additionally, the intense flashes of light produced by sprite lightning can hinder satellite and ground-based observations of the atmosphere. This interference can be particularly problematic for atmospheric science research using remote sensing instruments. Scientists are working to better understand the physical processes that underlie sprite lightning, as well as to develop strategies to mitigate its impact on radio communications and atmospheric research.

While there is no scientific evidence to support the idea that sprite lightning can be used as a gateway to another dimension, there have been theories proposed that suggest otherwise. In fact, some individuals have speculated that the intense electromagnetic energy produced by sprite lightning could be powerful enough to create a wormhole or a portal to another world.

There are still many mysteries surrounding sprite lightning, and the exact mechanisms behind this phenomenon are not well understood. However, it is known is sprite lightning involves a release of energy that is orders of magnitude greater than that produced by a standard lightning bolt. This energy release could theoretically generate a powerful shockwave that could rip open a portal to another dimension.

While this may sound like science fiction, it iss important to note that science has not ruled out the possibility of interdimensional travel. Theoretical physicists have proposed the existence of extra dimensions, beyond the familiar spatial dimensions and time dimensions that one experiences in their everyday life. It is possible that sprite lightning could somehow tap into these extra dimensions, allowing humans to travel to other worlds.

Second we have a theory that there are some sort of spectrum issues around the environment where Sasquatch lives. These dense spectrum issues could be a phenomenon that goes unseen by human eyes but exists in locations found by Sasquatch to be the most comfortable.

Some researchers believe Sasquatch may be drawn to certain locations with high electromagnetic activity. The idea of a perfect habitat for Sasquatch would therefore include an area with ample foliage and trees, minimal human interference and a higher-than-normal amount of electromagnetic energy. More research is needed to determine whether such a setting might create ideal conditions for these strange creatures to thrive while remaining hidden from view.

Our expedition ran out of time but we noticed that we needed to conduct more research on the environments where sasquatch sightings occur. To that end, we need to analyze the area to include other data along the electromagnetic spectrum. Exploring these "outside-the-box" possibilities will be an integral thread in vetting the validity of reported encounters. Once established, Sasquatch presence may lead to increased conservation efforts that benefit many other endangered species.

5

Environmental Reconstruction

The advancement of remote sensing technology and other multispectral technologies has opened the door to a world of possibilities for environmental reconstruction and saving endangered species. Remote sensing technologies such as near-infrared, infrared, and multispectral imaging can be used to detect signs of life in areas that would otherwise be difficult or impossible to access. These systems have already been employed with great success in various conservation efforts around the world, and could potentially provide invaluable insights into saving Sasquatch from extinction.

Here are a few of these advances in technology that could help in the discovery of Sasquatch:

1. LiDAR (Light Detection and Ranging): This technology uses laser pulses to measure the distance from a sensor to a target, allowing for 3D modeling of land surfaces, vegetation, and other objects. It is useful for high-resolution mapping of terrain and monitoring of environmental changes like deforestation or coastal erosion.
2. UAVs (Unmanned Aerial Vehicles): These small aircraft can fly over an area and collect data on the environment. They are used for surveying, monitoring climate change, assessing disaster damage, and tracking wildlife populations. Additionally, when equipped with the proper sensor, it can determine the health of meadows, wooded areas and wetlands.
3. GIS (Geographic Information Systems): This technology uses layers of geographic information to analyze patterns, trends, and relationships in the environment. It is used for a variety of purposes including risk assessment, land-use planning, resource management, and environmental impact studies.
4. Remote Sensing: This technology uses sensors to collect data from

satellites or aircrafts which can then be analyzed to gain insight into changes in the environment over time. It is used for a variety of applications including studying land cover, monitoring forest health, and mapping the ocean.
5. Sensors: These devices measure environmental conditions such as temperature, humidity, air quality, soil moisture, and more. They are used to monitor changes in the environment and can provide early warning of potential hazards like flooding or hazardous materials spills. They are also used for research purposes such as measuring the effects of climate change on wildlife populations or weather patterns.

Innovative Remote Sensing Technologies

Advances in infrared technology have revolutionized the way that heat is located and measured in the world. This technology involves using infrared cameras and sensors to detect the different wavelengths of light that are emitted from objects with varying temperatures. By measuring these wavelengths, it is possible to determine exactly how hot or cold an object is and track changes in temperature over time.

During the early 20th century, advances in photographic technology allowed for more sophisticated images of entire landscapes. In 1935, the first infrared photography was used to map agricultural areas. Around this same time, electromagnetic radiation was first used to measure objects on the Earth's surface. This technology, called radar, allowed for the detection of surface changes and other features on a global scale. By the early 1940s, world governments were actively utilizing this new technology during World War II to gain intelligence on enemy locations and movements.

The 1960s saw a dramatic expansion of remote sensing technology, with the first satellite images being captured by the Explorer 6 in 1959. In 1972, NASA launched the Landsat program which allowed for high-resolution images of Earth's surface to be taken from space.

The Landsat program is an ongoing mission by the United States Geological Survey (USGS) to collect images of Earth's surface from space. Begun in 1972, Landsat satellites have mapped the planet for more than four decades, providing valuable information about land cover and changes over time.

Infrared technology has proven to be an invaluable tool for finding animals in the wild. By tracking the infrared heat signatures of animals, it has become much easier to identify and track them over time. This is especially useful for conservation research, as it allows researchers to get a better understanding of animal behavior. Additionally, infrared technology can be used to identify and locate animals in the dark, making it a highly useful tool for nighttime tracking. Finally, infrared cameras can be used to detect heat signatures of injured or sick animals, allowing them to be treated quickly and humanely. Overall, infrared technology has proven to be an invaluable resource for animal conservation research and finding lost or injured animals.

One particularly useful component of this data is the near infrared (NIR) band, which measures reflected sunlight in the 800 to 900 nanometer range. These images are unique because they allow measurement of the health and vigor of vegetation from an aerial vantage point, making them an invaluable tool to manage Earth's resources. In addition, NIR images are also useful for monitoring land cover changes and identifying water bodies, soils, rocks and snow. This data can help better understand the dynamics of ecosystems, from photosynthesis to carbon uptake.

How people traditionally looked for Sasquatch

In order to find evidence of Sasquatch, people traditionally relied on eyewitness sightings and tracks in the wilderness. People often look for signs such as large footprints, partially eaten plants, ripped-up tree branches, or strange scat. Tracking Sasquatch through these signs can help paint a picture of the species' behavior and range.

Another traditional way for finding Sasquatch is to look for signs of their movement through the wilderness. This can include looking for paths that are wider than usual or broken vegetation in places where a creature might have passed through. People may also pay attention to unusual sounds, such as loud screams or knocks, which could be evidence of a Sasquatch nearby.

Finally, people can look for signs of Sasquatch activity by examining the droppings they leave behind. People often collect these droppings and send them to laboratories to be analyzed in order to gain more information about the species' health and diet. They may also look for unusual markings on trees that could indicate a Sasquatch has been nearby.

Unfortunately, due to the elusive nature of Sasquatch, these methods are often unreliable or inconclusive. This has made it increasingly difficult for researchers to find reliable evidence of its existence. Fortunately, new remote sensing technologies can help with environmental reconstruction, which could potentially aid in the search for Sasquatch.

Detecting Potential Habitats for Sasquatch

Remote sensing has the potential to provide valuable insight into the habitats and activities of Sasquatch. By using drone and satellite imagery along with other remote sensing techniques, researchers can map out areas that could be suitable for Sasquatch to live in.

Aerial imagery can help identify potential locations where dense forest cover, steep terrain, and large open areas exist. This type of habitat is typically sought out by Sasquatch, as it provides a good balance between concealment and access to food sources. A combination of multispectral and thermal imaging can also be used to map out potential habitats based on the vegetation structure and amount of sunlight in an area.

By using high-resolution optical and radar satellite imagery, researchers can map out slopes, streams, and other terrain features that may be suitable for Sasquatch. This type of data helps researchers gain an understanding of the topography and vegetation structure of an area, which can provide insight into the potential habitats that Sasquatch may inhabit.

In addition to using remote sensing techniques to identify potential habitats for Sasquatch, researchers have also used these methods in an attempt to gain an understanding of Sasquatch behavior. Thermal and other types of imagery can be used to identify areas where the Sasquatch are active, as well as to monitor their movements over time. This type of data can help researchers gain an understanding of the behavior and habitat preferences of Sasquatch, which can be used to further refine habitat models.

Use of technology for endangered species searches

One example of remote sensing technology success in the search for endangered species is the use of infrared imaging to locate and monitor African elephant populations. In 2013, scientists from the University of Twente in the Netherlands used airborne infrared systems to identify elephant populations across large areas of Central Africa and potential threats posed by poaching and habitat destruction. The results of their research showed that the elephants had declined in population numbers by an estimated 20-50% over the past decade.

A second example of a successful endangered species search is the use of satellite images to locate and monitor whale populations. Researchers from The University of Washington used high-resolution imagery to identify humpback whales in the northern Bering Sea, providing critical insight into the whales' movements and behaviors. This long-term data has been used to better understand whale populations around the world and help protect them from human-induced threats.

A third example is the use of Light Detection and Ranging (LIDAR) technology for locating and surveying bird species, particularly those that inhabit smaller habitats or are otherwise difficult to observe. Researchers have used lidar imagery to map bird populations in the California Central Valley, which has proven successful for monitoring migratory and threatened species such as Sandhill Cranes and Bald Eagles.

Benefits of Remote Sensing

Remote sensing technology has revolutionized the way endangered species are studied and monitored. This advanced technology allows us to observe animals from a safe distance, meaning no direct contact is required. This can be incredibly useful for surveying and tracking an animal's habitat without having to disrupt or disturb it. Additionally, aerial observation can provide us with valuable data in areas that are often difficult to access, such as mountainous terrain and dense forests.

Using advanced sensing technology also eliminates the need for human presence in dangerous areas or where access is limited. This means researchers can survey an area without putting themselves at risk of harm or danger. Additionally, aerial and satellite images can be used to detect environmental changes, allowing us to detect and monitor changes in an animal's habitat.

Finally, remote sensing technology can provide detailed data that would be impossible to obtain with traditional methods of searching for endangered species. For example, multispectral imaging can identify the presence of certain vegetation, water sources or minerals which may be critical habitats for species survival. This kind of data can be incredibly valuable for conservation and environmental reconstruction efforts by providing us with detailed data that would otherwise be impossible to obtain while also eliminating the need for direct contact or human presence in dangerous areas. Remote sensing technology is an invaluable tool for Sasquatch conservation and environmental reconstruction efforts.

6

Monitoring Changes in the Environment

Meaningful interpretation of remote sensing data allows for identification of changes in land cover, vegetation growth and water bodies over time. This information can be used to monitor environmental impacts associated with human activities such as deforestation, urbanization or climate change. Remote sensing also has applications for wildlife conservation. By monitoring changes in the environment, scientists can identify areas where Sasquatch may be present and better understand their habitat requirements. This data can then be used to create conservation plans or policies that will protect these species from harm.

Tracking polar bear populations using remote sensing technology has become an increasingly important research topic. This technology uses satellite imagery and other non-invasive techniques to study the habitats and behaviors of polar bears in their natural environment. The data collected helps researchers better understand how population trends are responding to climate change, pollution, and other environmental pressures.

One research project that has used remote sensing to study polar bear populations is the Arctic Remote Sensing Network (ARSN). This project is a collaborative effort between the University of Alaska Fairbanks, NASA Goddard Space Flight Center, and several other universities and organizations. ARSN focuses on collecting data from satellite imagery of areas along the coastlines of Canada and Alaska. This data is used to map out sea ice, vegetation and polar bear distributions. The project also uses GPS collars to track the movements of individual bears, which provides further insight into their behaviors and populations. With this information, researchers can better understand how these animals are responding to environmental changes in the Arctic Ocean ecosystem.

Research on monitoring animal migration paths with satellite imaging has grown in popularity over recent years. The use of satellite imaging to monitor and track the movements of animals across large distances brings an unprecedented level of information to conservationists, allowing them to better understand the behavior of threatened species and enact policies to help preserve them.

A specific research project in this area is the Monitor Migration in Africa (MMA) project, which seeks to understand how African migratory species respond to environmental change by using satellite imagery and other tracking systems. The MMA project uses satellite images to map out migration routes of some of the continent's most iconic wildlife such as zebras, elephants, and wildebeests. The project also uses drones to gather real-time data on the movements of animals, and logs their behavior over time in order to better predict the potential impacts of climate change on these species. By monitoring migration patterns closely and gathering data over a long period of time, this research can provide insight into how migratory species are adapting or struggling in response to changing environments.

Overall, the MMA project is a great example of how satellite imaging can be used to monitor Sasquatch migration paths in order to gain insight into wildlife behavior and its impact on conservation efforts. The findings from this project will be important for informing policies that help preserve these species and their habitats around the world.

A research project conducted by the Department of Wildlife Ecology and Conservation at the University of Florida focused on using aerial surveys to monitor wildlife habitat suitability. The project utilized a combination of UAS (unmanned aerial system) imagery and GIS (geographic information systems) analysis to assess the habitat resources for several species of Florida's wildlife.

The research team utilized a series of UAS flights to capture imagery throughout the study area. These images were then processed and used to generate digital elevation models (DEMs) that accurately depict terrain features such as topography, land cover and water bodies. The DEMs were then combined with GIS data sets on wildlife habitat distributions and species occurrence records to create models of habitat suitability for each species.

The results of this research project demonstrated that aerial surveys are an effective way to monitor wildlife habitat suitability on a regional scale. The team was able to identify areas of suitable habitat for several species of wildlife, and to suggest management strategies aimed at increasing the availability of suitable habitats in the study area. This research has implications for conservation across the state of Florida, as well as other regions with similar climates and habitats. Such studies can help to inform conservation decisions that protect wildlife in a cost-effective manner. In addition, this research demonstrates how UAS technology can provide invaluable information about habitat suitability for individual species and for entire species assemblages. This is a valuable tool for researchers and land managers in their efforts to conserve wildlife and ensure healthy ecosystems.

Monitoring environmental changes

Monitoring environmental changes is essential for identifying potential habitats for Sasquatch. The change in the environment has been a major factor in the shift of suitable habitat areas, and it is important to capture this data over time to effectively identify areas of potential habitation. In addition to this change, natural population fluctuations in wildlife due to predation or seasonal changes can significantly alter a habitat and thus require monitoring to ensure that suitable areas are identified. To find suitable habitats for Sasquatch, one can monitor different environmental parameters such as climate, vegetation, soil composition and available water sources. Additionally, understanding how changes in land use or human activity may impact potential habitats can help determine the best conservation strategies for preserving these habitats. In this way, environmental monitoring plays an important role in understanding and protecting potential Sasquatch habitats.

Researchers can gain a better understanding of how human activity is impacting potential habitats for Sasquatch and develop strategies to protect them from further damage. This knowledge can then be used to preserve and restore these areas so that they remain viable for future generations of Sasquatch. Ongoing monitoring is essential to ensure suitable habitat identification and preservation so that Sasquatch may live on for many years to come

Suitable habitats for species reintroduction

Restoring the presence of endangered and threatened species has directly benefitted from the use of remote sensing input. For instance, multispectral satellite imagery can be used to identify land cover type and vegetation structure, which are important factors in determining suitable habitats for reintroduction. This data can be used to monitor changes in vegetation structure over time due to various human activities, including urbanization and deforestation. By analyzing these trends, conservation managers can assess the impact of these activities on species reintroduction and develop strategies to mitigate them. Ultimately, remote sensing data can provide scientists and conservation managers with the information they need to ensure successful reintroductions.

A great example of how this can be done comes from a study conducted by researchers at the University of Florida on the reintroduction of Florida panthers (Puma concolor coryi). The team used remotely sensed data to identify potential suitable habitats for panther reintroduction, by combining topographic LiDAR data with land cover information from satellite imagery. They then created a habitat suitability model to prioritize areas for reintroduction, and used genetic data to identify the best sources of panthers for reintroduction. Finally, they monitored the development of the new population over time using camera traps. The study found that remote sensing data was key in determining suitable habitats for panther reintroduction and in tracking the success of their project. It also demonstrated that combining remote sensing data with genetic data can provide powerful insights into species reintroduction programs, allowing for better selection of sources, more accurate prediction of population growth, and more effective management strategies. This study highlights the potential of using remote sensing data to successfully plan and monitor species reintroduction programs.

Another example of how remote sensing data has been used to identify areas for species reintroduction can be seen in a study conducted on the reintroduction of red wolves (Canis rufus). In this study, researchers from North Carolina State University and the U.S. Fish and Wildlife Service used satellite imagery and geospatial analysis techniques to assess the suitability of potential reintroduction sites. They then combined this data with information on land cover, topography, and habitat use to develop a habitat suitability model. This model was used to identify areas that could provide suitable habitats for red wolf reintroduction. The study also incorporated genetic data from existing wolves in order to select optimal locations for releasing wolves. This study demonstrates how using remote sensing data in combination with genetic data can be beneficial for species reintroduction programs, providing scientists and conservation managers with the information they need to ensure successful reintroductions.

What it means to our Sasquatch research

Remote sensing data can be an invaluable tool for conservation and species reintroduction efforts. It allows researchers to track environmental changes over time, assess the suitability of potential habitats for reintroductions, and gain insights into human activities that may be impacting the environment. In our Sasquatch research specifically, this type of technology could provide us with valuable information about their habitat requirements and how one can best ensure their long-term survival. With more knowledge on these topics comes greater ability to protect them from extinction in the future. Overall, remote sensing data can be an important part of the Sasquatch conservation effort. It is our hope that this technology will enable us to further understand and protect this unique species. Doing so can ensure they remain a part of the world for many years to come.

7

Strategies for Restoring Habitat

Restoring and maintaining healthy ecosystems is paramount for ensuring the continued existence of Sasquatch. Without appropriate habitat preservation, these populations may not thrive and are at risk of extinction.In order to create an ideal environment for these creatures, it is important to understand how to restore and improve habitats that support plant and animal life. These efforts should include consideration of soil quality, biodiversity, managing invasive species and restoring areas adjacent to streams and rivers (riparian areas).

Soil quality is important for the environment, but can quickly become compromised when allowed to degrade. This degradation is largely due to runoff pollution from both agricultural and urban sources. In the agricultural setting, creation of buffers of grass along streams and rivers has been shown to reduce runoff pollutants by soaking up and processing the substances, thus preventing them from reaching the water. In the urban setting, catch basins and reservoirs collect storm runoff along with all the contaminants caught up in that flow, and once collected can either be skimmed off or filtered. Reducing this contamination can help restore balance by increasing plant diversity and wildlife habitats.

Increasing biodiversity is important to an ecosystem by maintaining its balance and health. The complex interaction of diverse populations creates a strong, resilient system that can better adapt to changing environmental conditions. This also provides greater food resources for other organisms, improved water quality, cleaner air, and more varied recreational activities for humans. It can even make the area more resistant to natural disasters like floods or fires. Biodiversity increases the overall health of the environment, making it easier for all creatures, including Sasquatch, to survive and thrive.

Managing invasive species is important to an ecosystem because they can have a negative impact on the environment. Invasive species spread disease and out-compete native plants and animals, thereby reducing biodiversity. Controlling populations of these ecologic invaders helps protect habitat health and maintain the integrity of an ecosystem's natural resources.

Riparian areas are essential in maintaining soil health, as they help filter pollutants and provide habitats for water-dependent species. These ecosystems also play an important role in flood control by slowing down the flow of water and reducing soil erosion. The roots of riparian vegetation help to retain soil and prevent erosion, while the trees and shrubs provide shade for the streambanks and cool water temperatures. Restoring these areas can help to reduce the impacts of flooding, improve water quality, and create healthy habitats for fish and other aquatic organisms. This helps to create an ideal environment for beneficial insects, birds, turtles, amphibians, and other species that live in these ecosystems.

Strategies to restore habitats

Adding native plants back into the ecosystem serves to recondition habitats. This can be done through planting, seeding, or transplanting efforts. Because native plants often have deeper root systems, this helps to prevent soil erosion and can provide food and shelter for wildlife.

Drones are becoming increasingly popular for seeding native plants back into an ecosystem. A drone-based seeding system works by using a small, autonomous unmanned aerial vehicle (UAV) to spread seeds over a larger area than would be possible with traditional methods. The drone can be programmed to fly in predetermined patterns or it can even be outfitted with image recognition software that allows it to detect areas where the plants could flourish. Drone-based seeding systems are becoming increasingly popular because they provide a cost-effective and efficient way to restore and improve habitats. Additionally, they can help minimize human contact in fragile ecosystems, reducing potential damage and disruption of existing habitats.

Incorporating water sources can be an important way to restore and improve habitats. This could involve ponds or streams, existing waterways, or even rain gardens that collect runoff from buildings and concrete surfaces. Not only do these methods provide water for local wildlife, they also help to reduce flooding and recharge ground water. Once potential water sources have been identified, it is important to consider the impacts of any changes made in order to ensure that local ecosystems remain healthy and balanced. For example, if rerouting existing waterways is necessary, precautions should be taken to protect any fish or wildlife living in the area. Additionally, when installing artificial sources such as rain gardens or ponds, it is important to use plants that are native to the region and avoid introducing invasive species.

Identifying pollution levels in a habitat is one of the keys to restoring and improving it. When assessing the environmental conditions of a habitat, a number of different pollutants need to be taken into account. These include air and water contaminants such as heavy metals, nitrogen, and phosphorous; chemical pollutants such as pesticides and herbicides; as well as light and noise pollution from nearby urban areas.

Once the sources and levels of pollution have been identified, it is important to determine the best way to fix it. For example, if a habitat contains high levels of air pollutants such as nitrogen dioxide or carbon monoxide, measures need to be taken to reduce these levels in order to improve air quality. This could involve installing industrial air filters or working with local industries to reduce the amount of pollutants they are emitting. Similarly, if a habitat contains high levels of agricultural runoff, measures need to be taken to prevent this from happening in the future.

Overall, restoring and improving habitats requires a comprehensive approach that addresses all of the environmental issues present in a given area. This includes creating water sources, reducing pollution levels, and controlling human contact with fragile ecosystems. When done properly, these efforts can have a significant impact on local wildlife populations and help create healthier, more resilient habitats.

Examples of successful habitat restoration

Bald eagles had reached dangerously low numbers, leading to their placement on the Endangered Species List in 1978. Through collaborative efforts between state, federal, and private organizations, a robust restoration program was put in place and by 2007, the bald eagle was removed from the Endangered Species List.

The restoration of bald eagle habitats involved protecting nesting areas from disturbance, such as human interference and predation. This included creating buffer zones around nests to reduce the risk of habitat destruction. Nesting platforms were also set up in appropriate locations for eagles to nest on, providing them with more protection from predators and weather conditions. Finally, artificial nest structures were created by volunteers to provide additional sites for bald eagles to lay their eggs safely without fear of predation or disruption. By taking these steps, the number of suitable nesting sites increased dramatically which led to a significant increase in the population size of bald eagles across North America.

Creating an ideal Sasquatch environment

Creating a desirable habitat for these creatures involves developing resources to sustain and promote the growth of plant and animal life known to allow them to flourish. When these resources are available in abundance, it creates a healthy balance between species which helps to promote biodiversity and keep the ecosystem in equilibrium. Furthermore, a healthy environment can lead to increased abundance of Sasquatch, which can then contribute to the ecological balance of an area. A balanced ecosystem is beneficial because it helps to reduce the impacts of disease and environmental degradation, and provides numerous other benefits such as clean air and water. Creating an ideal environment for Sasquatch will help to ensure that the species has a chance at survival and can continue to contribute to healthy ecosystems.

Reestablishing a habitat in which Sasquatch can thrive could allow them to feel safe and allow for researchers to actually observe them in the wild. By providing an environment that is safe, rich in resources, and suitable for breeding, it would be possible for Sasquatch to establish a population that could be studied.

When evidence of Sasquatch's existence comes to fruition, this could lead to placement onto the endangered species list due to their low population numbers. With a lack of data regarding their numbers and habitats, it is difficult to determine the exact level of threat they face. However, if researchers are able to observe them in their natural habitat and document their behaviors, this could provide valuable information to help assess their risk of extinction. This could potentially lead to the development of conservation plans as well as legal protection that could ensure the survival of Sasquatch in the future. Additionally, if they are placed on the endangered species list, it would open up more funding for research and allow for better protection of this mysterious creature. With increased knowledge and protection, it could lead to the preservation of Sasquatch in the wild for future generations.

8

Potential Challenges

Reestablishing an environment where Sasquatch can thrive is no easy feat. There are a number of potential challenges that need to be addressed in order for this species to make a successful comeback, such as limited funding and resources, human interference, and the difficulty of remote sensing environmental restoration strategies. With these obstacles in mind, it is important to take an informed and holistic approach when attempting to restore habitats for the Sasquatch population. By understanding the full scope of all potential issues involved with re-establishing viable ecosystems, one can develop effective solutions that work toward ensuring their long-term survival.

Limited Resources

Creating a good habitat for Sasquatch would require resources that could provide protection from things like extreme weather while allowing enough space for the creature to roam and have access to food sources. Depending on the location, these resources may include trees or other vegetation to create shelter, materials such as logs or rocks to build dens or hideouts, and open spaces of land to roam and feed. Water sources such as ponds or streams, would also be a necessity.

Areas with bountiful vegetation are ideal for providing these resources in ample supply, but can be rare or hard to get due to a variety of factors. Pristine water sources may be difficult to locate, and are becoming increasingly rare because of deforestation or development projects that disrupt natural ecosystems. Finding an area with adequate vegetation can prove challenging if human activity has already encroached into Sasquatch living space, making it more difficult for Sasquatch habitats to exist naturally in many parts of the world.

In order to create an ideal habitat for Sasquatch, it is essential to protect the existing resources. This includes conserving and preserving the land, water, air, and wildlife that live there by implementing such strategies as creating wildlife corridors, limiting human access to certain areas, and restoring damaged habitats. In doing so, these efforts can ensure that the resources available will be able to support Sasquatch populations in the future. Additionally, conservation work in these areas can also help to safeguard the well-being of other species living in these habitats.

Limited Funding

Regenerating old-growth forests and previously established transition zones is a monumental mission, with funding for such projects in the range of $5000-$150,000 per acre depending on the scope and size of the project. This may include costs associated with soil excavation and transport, planting native vegetation, acquiring the necessary permits and approvals, and any additional expense associated with the project. Additionally, these costs may be even higher in remote areas where access is limited or resources are scarce. Considering it can take many years for a wetland to fully restore itself; funding must be provided over an extended period of time to ensure successful restoration of the habitat. With limited funding sources, it can be difficult to secure the necessary capital for these large-scale projects.

To procure the necessary funds, donations from corporate, private and philanthropic sources must be considered. These donors may be willing to contribute to environmental conservation initiatives due to the potential positive publicity associated with such projects. In fact, many donors are motivated by the desire to help protect and preserve the natural environment, and may be willing to donate funds for Sasquatch conservation initiatives.

Grants from the government or other organizations would be another source of funding, as well as environmental subsidies and loans from federal agencies such as the Department of the Interior or Environmental Protection Agency. Similarly, there are local, state, and private foundations that offer grant opportunities for wildlife conservation initiatives.

Donations from individuals can be obtained through a variety of platforms such as social media, Kickstarter campaigns, and traditional fundraising efforts. People may also be willing to contribute in the form of volunteer hours or materials donations. By creating an awareness campaign to inform the public about the need for Sasquatch conservation initiatives and what they can do to help, environmental restoration funds may be obtained more easily.

Human interference

As humans are increasingly encroaching on natural habitats where Sasquatch are thought to be, they could create a disruption in the environment that Sasquatch needs to survive, such as that seen with the use of DDT during the mid-20th century. DDT was used extensively to combat insect-borne diseases such as malaria, but it also caused significant environmental damage. The pesticide was linked to the decline of several species of birds, including bald eagles, peregrine falcons, and brown pelicans, who either ingested the toxic chemical or consumed contaminated prey. DDT also caused eggshell thinning in many bird species which affected their ability to reproduce, leading to a steep decline in population numbers. These effects have been reversed since the ban on DDT in 1972, but it serves as an example of the far-reaching consequences such interference can have on wildlife and ecosystems.

Habitat deforestation for farming or industrial development has been an ongoing problem since ancient times, when humans first began clearing forests for agricultural purposes. In modern times, deforestation has had a devastating effect on many species, such as was seen with the Giant Anteater as a result of destruction of the rainforest in the Amazon.

In California, the Channel Islands were once home to one of the largest colonies of seabirds in the world, with an estimated 5 million birds living on them. However, over the years, urbanization and development of the islands have destroyed much of their natural habitat and caused a decline in seabird populations. This has had devastating effects on species such as the Western Gull, which is now considered endangered. These changes to the environment have also impacted other wildlife, including seals, otters and whales.

Fisheries have led to the depletion of specific aquatic populations, as well as damage to their habitats and ecosystems. This was seen with the decline of Atlantic cod in the waters off Newfoundland, Canada, in the late 20th century. This began when foreign fishing fleets expanded their operations to take advantage of newfound fish stocks, and soon after, industrial-scale cod fishing began. The combination of heavy fishing pressure and an increase in local demand resulted in overfishing, leading to a collapse of the cod population. As a result, thousands of people were left jobless and many traditional fishing communities suffered.

In the 19th century, bison were hunted for their hides and meat at unprecedented levels, leading to their extinction from much of the plains. In addition, the introduction of cattle to the plains destroyed bison habitat and competition for forage increased further. As a result, millions of bison were killed in a short time period, resulting in their drastic population decrease and eventual extinction from much of North America.

The introduction of non-native species into an ecosystem, can disrupt the balance of nature and cause serious damage to wildlife and habitats. This includes animals such as rats, mice, rabbits and deer that can spread disease or compete with native species for food and resources. In New Zealand in the mid-1800s, the British had occupied the country and began introducing rabbits, hares, and other animals for hunting purposes. As a result, these species quickly spread throughout the islands and disrupted the natural balance of the environment. Their population growth caused serious damage to the native flora and fauna, as well as other habitats such as wetlands and grasslands. This eventually led to the near-extinction of some native species and a decline in biodiversity throughout New Zealand.

Remote Sensing

Remote sensing environmental restoration strategies are complex processes that involve a variety of technical components such as software and hardware. Implementing these strategies can be difficult without the right resources and personnel, as errors in implementation can have significant consequences.

Remote sensing environmental restoration strategies often require the use of sophisticated software, such as geographic information systems (GIS), to accurately analyze and track changes in the landscape over time. Additionally, hardware components may be necessary to collect data from the environment, such as satellites and unmanned aerial vehicles (UAVs). Properly implementing these strategies involves a significant amount of planning, as well as the expertise of personnel skilled in environmental issues and remote sensing technologies. It is also important to consider the potential effects of errors; a poorly implemented restoration strategy may result in unexpected consequences for the environment.

Multiple Stakeholders and Politics

Environmental restoration projects often involve multiple stakeholders with different interests that must be taken into consideration when making decisions. This can lead to disagreements among stakeholders that can be challenging to navigate, resulting in costly delays in implementation.

When it comes to environmental restoration, politics can often play a major role in the decision-making process. While stakeholders may have the same goal of restoring the environment, they can disagree on the best way to achieve that goal. This can lead to debates over which methods should be used and how much money should be allocated for various projects. Additionally, political leaders may have their own agendas that can influence the decisions made, or they may be influenced by lobbyists who support certain solutions. As a result, the process of environmental restoration can often be complicated and difficult to navigate. It is important for stakeholders to remain open-minded and consider all perspectives in order to ensure that environmental restoration projects are successful.

Restoring the Sasquatch species through remote sensing environmental restoration strategies can be a difficult task due to limited funding and resources as well as human interference. However, with careful planning and implementation of proper protocols, it is possible to achieve successful conservation outcomes for this iconic species. To ensure success in these efforts, stakeholders should consider investing in research partnerships between scientists and local communities to gain insight into how best to utilize existing tools while also exploring new methods that may prove beneficial. Through collaboration on both regional and global scales, a better future for all wildlife populations can be created, including the beloved Sasquatch.

Benefits of Restoration Efforts

Restoration efforts are becoming increasingly important for plants, animals and the overall health of the ecocommunity, including human health. Bringing back natural habitats and ecosystems leads to cleaner air and water, beneficial to wildlife. Mindfulness of these activities is bringing awareness to the plight of Sasquatch and his disappearing environment. This reversal will bring about changes for all members of the ecocommunity.

Increased Biodiversity

When biodiversity increases, the health of local communities is improved in a number of ways. A variety of food sources provides a better chance at proper nutrition, which is essential. A wide range of plant and animal species boosts habitats for other organisms and provides shelter from the elements. Studies have shown that areas with greater biodiversity are more resilient to climate change, making them better prepared for any storm or disaster.

A study conducted by the United States Geological Survey (USGS) found that areas with greater species diversity were better able to withstand extreme weather events than those with fewer species. The USGS study focused on the effect of extreme weather events, such as floods, droughts, and hurricanes, on species diversity. The researchers evaluated over 100 sites across the United States and found a clear correlation between increased species diversity and increased resilience to these weather events. For example, areas with high levels of plant species richness were more resistant to flooding than those with low species richness. Similarly, areas with greater bird species abundance were better able to cope with drought conditions than sites with fewer species. The researchers concluded that greater species diversity leads to a more resilient ecosystem, one that is better equipped to withstand extreme weather events.

Another example of a study in biodiversity was conducted in 2020 by researchers from the University of California, Davis. This study explored the effects of higher levels of species diversity on carbon storage and sequestration. The researchers found that areas with greater species diversity had more carbon stored in their soils than sites with fewer species. In addition, they also discovered that sites with greater species richness had higher levels of soil nutrient uptake, which can lead to healthier soils and increased crop productivity. The researchers concluded that increasing biodiversity can lead to a more efficient carbon cycle, resulting in the mitigation of climate change. Furthermore, they suggested that restoring or conserving areas with high species diversity could be an important tool for tackling climate change. Thus, this study revealed the importance of biodiversity in carbon storage and sequestration.

Growth of endangered species populations

Creating an appropriate environment can promote the growth of endangered species populations that may previously have been at risk of extinction. Restorative efforts in an at-risk environment provides endangered species with the chance to thrive by creating a safe and sustainable habitat. With the right resources, endangered species can be properly nurtured and cared for, allowing them to reproduce and grow without the threat of poaching or the destruction of their habitats.

One historical example of endangered species growth that has benefitted humans is the California Condor. Following a drastic reduction in population due to hunting and poaching, conservationists began to take action in the 1980s through captive breeding programs. These efforts were successful, leading to a dramatic increase in the population of California condors from 22 individuals in 1982 to 436 as of 2020. This growth has allowed humans to benefit from watching these magnificent birds in their natural habitats and has helped to preserve the diverse genetic heritage associated with this species.

If the Sasquatch population were to grow, it would have a positive impact on humans in a number of ways. It would bring attention to the need for conservation of these mysterious creatures, increasing awareness of the threats they face from habitat destruction. An increase in their numbers would also allow scientists an opportunity to study this elusive species, possibly leading to a greater understanding of their behavior and ecology.

Improved Economic Opportunities

Overall, environmental restoration offers a variety of economic benefits that can help local communities thrive in a sustainable way. From creating jobs to reducing the costs associated with flood damage and health care, restoring the environment can advance economic opportunity while preserving the remote habitat preferred by the Sasquatch species.

Because Sasquatch has a global domain, restoring and protecting natural resources can have a positive effect on global markets. For instance, increasing biodiversity by planting trees or other vegetation helps to reduce emissions of planet-warming greenhouse gases. This, in turn, reduces the prevalence of climate change and its associated costs. Similarly, preserving water resources helps to reduce water scarcity and its associated economic impacts. By doing so, global markets are provided with a more stable and secure foundation for long-term growth.

Ultimately, environmental restoration is essential for creating improved economic opportunities in local communities around the world. These benefits can be felt on both a micro and macro level, helping to create jobs, reduce costs associated with climate change and water scarcity, and provide a stable foundation for long-term global economic growth.

Positive Impact on Human Health

Environment restoration efforts can lead to improved air and water quality, providing a healthier and more sustainable environment for humans to live in. An example this is the clean-up of the Hudson River in New York. In the early 20th century, unchecked industrial and commercial waste had severely polluted the Hudson River, posing a health hazard to those living near it. After decades of legal battles, public outcry, and advocacy work from environmental organizations, the Hudson River was finally cleaned up in the 1980s. This led to improved water quality, and a thriving fish population that supported local fishermen. Today, the river is an important source of recreation and a symbol of environmental success. Its restoration is a testament to the importance of environmental conservation and protection.

The health benefits of environmental restoration do not just apply to water quality - restoring forests can also lead to improved air quality. Trees absorb carbon dioxide from the atmosphere and release oxygen back into it, helping to reduce air pollution levels. The presence of healthy forests can also provide a natural refuge for numerous species, creating a more biodiverse ecosystem that is better able to support human life. In addition, the presence of green spaces and parks in urban areas can improve physical and mental health by providing places for people to relax and exercise.

Benefits to Sasquatch

Restoration efforts can have significant benefits for Sasquatch, as with other species. To create a more ideal living environment for the elusive creatures, increasing biodiversity and improving habitat health will be indispensable. A more robust ecosystem gives Sasquatch a wider variety of food sources and shelter options, allowing them to live and thrive in areas previously unable to support them. As a result, the improved health of plants and animals in the vicinity would provide Sasquatch with more prey and competitors, resulting in a healthier balance of resources.

In an ideal environment, Sasquatch typically seek out places that provide ample shelter and protection from the elements. They will inhabit densely wooded areas with large trees and thick vegetation, providing them with cover and a place to avoid potential predators. Sasquatch may also take advantage of natural shelters such as caves and rocky overhangs, allowing them to safely observe their surroundings and search for food sources without coming into contact with humans or other potential threats.

Sasquatch are omnivorous creatures, meaning they eat both plants and animals. They are particularly adept hunters in the wild and can consume a variety of prey. Sasquatch will also feed on fruits, berries, nuts and other vegetation when available. Some have even been known to roam close to human settlements in search of food scraps and other edible items. By consuming a range of foods, Sasquatch are able to remain well-nourished in their natural habitats.

One of the most famous sightings of a Sasquatch taking food scraps from humans occurred in 1951, when Canadian trapper George Constable encountered a mysterious creature near his cabin in the woods of British Columbia. According to Constable, he saw the Sasquatch near his home one evening and noticed it was eating some food scraps he had left out. Constable described the creature as being seven feet tall, stoutly built and covered in dark fur. He said that it had "wide shoulders, long arms, and a large head with a flat face". The Sasquatch then quickly vanished into the woods without making any sound. If the local ecosystem had been in better balance and provided a more varied range of food sources and shelter, it is likely that Sasquatch would not have needed to roam close to the trapper's cabin for food scraps. With access to a wider selection of wild prey and healthy vegetation, Sasquatch could adequately sustain themselves without coming into contact with humans or being seen by people.

    Restoring ecosystems can provide a buffer between wild animals and human activity, giving Sasquatch more space to roam freely. These efforts play an integral role in creating an environment in which the ecocummunity will thrive, thereby generating benefits for all. Recreating that natural balance and providing a wide range of food sources and shelter options creates an ideal living environment for all species. To ensure efforts are successful in preserving biodiversity and protecting wildlife habitats, it is important to take action now before further damage is done. Everyone must do their part in supporting sustainable development practices as well as encouraging conservation initiatives so that future generations may benefit from healthy ecosystems.

Sasquatch Conservation Efforts

As the debate over the existence of Bigfoot continues, conservation efforts for this mysterious creature are becoming more important. With increased public interest in Sasquatch sightings and reported encounters, it is essential to consider what strategies can be implemented to ensure their survival in the future. From habitat protection initiatives to educational campaigns, there is a range of potential approaches that could be employed to help protect this elusive species.

Moving Forward

Ongoing positive measures are making a difference in providing a sustainable homestead for these elusive creatures. An important example of a reforestation project is the restoration efforts of the American Chestnut tree in the U.S., which began in 1983. The American Chestnut was once one of the most widespread trees throughout eastern North America, and a primary source of food for wildlife such as squirrels and bears. Unfortunately, an invasive blight put the population of this species at great risk. Restoration efforts have since been successful in restoring much of the Chestnut's former habitat, providing a safe and secure environment for Sasquatch as well.

Old growth forestation provides a safe, secure environment for Sasquatch through a number of strategies. One of the most effective ways to do this is through improved logging practices. When logging takes place in a particular area, the trees and plants that provide cover and food for Sasquatch may be destroyed or altered. To avoid this, increased regulations and enforcement of restrictions for heavy equipment in these areas can help to ensure the safety of Sasquatch.

Logging can be done in a way that protects the species living in an old-growth forest by selecting trees to be cut based on their age, size and growth patterns. Additionally, rotational logging, a method where only some trees are removed at one time, can help to preserve the overall habitat structure, allowing for Sasquatch and other species to find shelter and food. Clearcutting should be avoided when possible, as it can lead to large biodiversity loss within a small area. Finally, leaving buffers around specific tree stands or concentrating harvests on certain areas of the forest instead of widespread clearcutting would help protect Sasquatch habitats from being over-harvested.

Mountainous regions of North America are particularly suited to Sasquatch, as they provide an abundance of natural resources. For example, the Cascade Range in Washington State provides an array of plant life which can provide cover and sustenance for these creatures. Likewise, the Rocky Mountains in Colorado and Canada offer a variety of hiding spots for Sasquatch which can also be used for protection. The low human density found in these areas makes them less likely to attract unwanted attention from interlopers inquiring about sightings or trying to connect with these creatures directly.

Human-Sasquatch Interaction

One of the most inspiring Sasquatch encounters occurred in 2002, when a family camping in the Blue Mountains of Oregon had an unexpected visitor. A female Sasquatch appeared from nearby woods and walked to the campfire area. She seemingly knew that no harm would come to her and so stayed for some time – even interacting with the family and accepting a gift of an apple. After some time, the Sasquatch returned to the woods, bringing the family's encounter to a peaceful end.

While positive interactions are encouraging, there are also those instances where things do not go as well. An example of this occurred in 2019 in the state of Washington. During the summer months, a family living near Mount St. Helens noticed strange sounds coming from the woods behind their house and began to investigate. After tracking the noises to a nearby ravine, they discovered a female Sasquatch and her two offspring. The family attempted to approach the creatures but were met with hostility and aggression from the Sasquatch, who growled and raised her arms in defense. Eventually, the family retreated from the area and has not seen the Sasquatch again.

The above examples demonstrate that humans and Sasquatch can coexist peacefully in the same environment, provided we are respectful and understanding toward these creatures. In the case of the Oregonians, a local family warmly welcomed a solitary Sasquatch into their campground with food and treats. However, when humans approach Sasquatch as was seen with the Mount St. Helens encounter, without proper caution and understanding, the results can be dangerous and lead to regrettable incidents. It is clear that to maintain continued coexistence with these mysterious creatures, one must first strive to understand them.

Reintroduction Programs

Focusing on how to safely and successfully reintroduce Sasquatch into areas where they have been absent for a long period is becoming increasingly important in Sasquatch conservation. These programs require careful planning, research, and monitoring in order to ensure the safety of participants and the Sasquatch themselves. In addition, introduction efforts should consider a variety of factors such as habitat conditions, available resources and potential risks. If these considerations are taken into account, reintroduction programs can be successful in restoring healthy Sasquatch populations to previously uninhabited areas. In fact, the organization "Saving Sasquatch" is dedicated to protecting the dwindling Sasquatch population through environmental conservation and reintroduction programs and have been working to protect habitats, monitor Sasquatch populations, and raise public awareness about the threats facing these creatures.

No Sasquatch has ever been captured, and it is highly unlikely that this will ever happen given the sheer difficulty of such a task. As such, any reintroduction of the species back into their native habitats must be done in an entirely different manner such as utilization of remote sensing technology to determine viable locations. This type of technology could help to map out areas with suitable terrain and vegetation that would provide a conducive environment for Sasquatch. Continued research into their diet, behavior, and habitat requirements could help inform conservation efforts on where to create protected habitats in order to support the species' reintroduction. This is what "Saving Sasquatch" is attempting to do with both its research and environmental restoration efforts.

Conservation Education and Awareness

Conservation education and awareness is an important concept taught in American schools. In today's world, it is vital that students understand the importance of preserving the environment and its resources. This can be done through a variety of activities both inside and outside of the classroom. Teachers can discuss environmental topics such as environmental changes, pollution, and the sustainable use of resources. Projects encourage students to think proactively about the environment and come up with possible solutions to environmental problems. Through hands-on activities such as field trips or community service projects, students gain a better understanding of their own roles in conserving natural resources.

In addition to the classroom instruction, these concepts can be taught through other resources such as books, videos, and online activities. Books geared toward children can teach them about different aspects of the environment, while videos may show how certain practices are damaging the environment. Online activities like interactive games or simulations can engage students in topics related to conservation and give them a better understanding of their own impact on the environment.

In an effort to provide education to the general public, the National Park Service has been creating awareness about feeding animals or bear attacks for decades. It began in the early 20th century when Yellowstone National Park implemented regulations prohibiting visitors from feeding wildlife. Since then, the park service has developed a variety of ways to educate visitors about the potential danger of feeding animals. One example is the Ranger-led program in Glacier National Park, which educates visitors on the proper way to interact with wildlife. The program also provides information about potential hazards such as bear attacks and what should be done if a visitor encounters an aggressive animal. In addition, the park service has implemented signage throughout parks to remind visitors to feed wildlife and that it is illegal in most parks. The park service has also created a variety of educational materials and videos to help visitors better understand the risks associated with feeding animals. By raising awareness about the dangers of feeding wildlife, the National Park Service is helping to ensure that visitors have safe and enjoyable experiences in national parks.

The conservation of Sasquatch habitats requires a heightened level of knowledge and awareness due to the unique nature of these environments. Like with the example of feeding wildlife, the National Park Service should develop methods for educating visitors about the importance of preserving Sasquatch habitats and the potential danger involved.

In parks where Sasquatch sightings have occurred, the park service should implement special regulations to ensure that visitors do not interfere with their habitats. Signs can be placed throughout the parks to inform visitors of these rules and to remind them of the importance of preserving Sasquatch environments. In addition, Ranger-led programs can be used as an educational tool to teach visitors about proper practices when exploring Sasquatch habitats. The park service should also create a variety of educational materials and videos to help visitors better understand the risks associated with interfering with these unique ecosystems.

"Saving Sasquatch" will provide educational materials that focus on what to do when encountering a Sasquatch. These materials will provide clear guidelines for how a person should respond if they come across one of these creatures in the wild, including advice on avoiding direct contact and remaining at a safe distance. The endeavors being undertaken to save Sasquatch are important in making sure the environment is enhanced. More detailed information is available at; savingsasquatch.org.

Does Sasquatch really exist?

This question remains unanswered; time and continued research will provide that information in due course. Whether he is intradimensional or extradimensional, or even has a link to the uberelectrical field generated by the atmospheric sprites, his presence will be verified at some point. One could even say that because no empirical evidence has been generated, it is obvious that he does not exist. However, the continually increasing number of sightings and encounters would tend to call that position into question.

References

Deinet S, Böhning-Gaese K, Gardner TA (2016). Global patterns of extinction risk for birds and mammals in the Anthropocene. Proc Natl Acad Sci USA 113(24): 6670–6675. https://doi.org/10.1073/pnas.1603783113

Johnson, K., Dunn, D., & Houghton, T. (2014). Human Impact on Wild Animal Populations in North America. Frontiers in Ecology and Evolution, 2(1), 1-12. doi:10.3389/fevo.2014.00001

Sarvotham, V., Al-Hamadani, A., & Mecikalski, J. (2020). Remote Sensing Technologies in the Context of Conservation of Endangered Species - a Review.

Sasquatch History (2022). Retrieved from https://savingsasquatch.org/history/

Scott, E., & Bendell, J. F. (2004). Human Impacts on Wild Animal Populations of the Sierra Madre Occidental: A Review of Contemporary Threats to Mexican Montane Forests and Their Wildlife Communities. Conservation Biology, 185(5), 1242-1252. doi:10.1111/j.1523-1739.2004.00087_1x

Strain, Katherine Moskowitz (2012). Mayak Datat: The Hairy Man Pictographs. The Relict Hominoid Inquiry 1:1-12. https://www.isu.edu/media/libraries/rhi/research-papers/Mayak-Datat-Hairy-Man-Pictographs-1.pdf

Acknowledgments

We would like to thank Ali Davison for her assistance with some of the content of this book, Rachelle Berggoetz for her expertise and editing skills, The State of Michigan Library for their timely suggestions and assistance, The State of Washington for their Sasquatch Sightings Data Base, The State of Oregon for their access and use of their supporting data and The US Geologic Survey Department for their elevation landscape mapping. Additional credit goes to the US Department of Agriculture and the US Forestry Service.

NOTE - As a point of clarification, the terms Bigfoot, Sasquatch and even Yeti, seem to be interchangeable, describing the same creature.

Editor's Note

For me, this project was a labor of love for my father, Ray David. When he told me what he was putting together, I'll be honest – I was pretty surprised. He's a multifaceted and interesting guy, don't get me wrong, but this was not a subject I'd have expected him to be interested in, let alone write a book about, let even more alone doing so after having just been diagnosed with terminal pancreatic cancer. The time we spent together combing through the text he provided taught me that 'impossible' is not really in his wheelhouse – whether that's referring to the bounds of the actual universe or the potential for surviving nearly 2 years (and counting!) into a diagnosis that was to have sent him to Heaven after a handful of months. So for anyone reading this, I ask that you raise a glass to this very ornery, funny, and intelligent man – Dad, try to keep at least 2 wheels on the road when you round the corner, and for the love of everything good, don't run over St. Peter when you get there!

ABOUT THE AUTHORS

Ray David
Ray did his undergraduate studies in chemistry and biology. He owned a materials testing laboratory where 80% of the testing was conducted for the Department of Defense. Currently he is the holder of three separate licenses with the Federal Aviation Agency (pilot, mechanics and non-destructive testing). He was then appointed the first Environmental Lab Director of the Annis Water Research Institute at Grand Valley State University. Ray then acquired an environmental testing and research facility. In this capacity they tested air, soil and water for the EPA Priority Pollutants in addition to consulting with both private industry and state and federal agencies.

Jaremy Latchaw
Jeremy is a leading drone and remote-sensing expert. He started working with autonomous vehicles in 2009 and started a drone company in 2016. He quickly grew into a sought-after speaker and subject matter expert in that industry.

In 2015 Jeremy and Ray combined their strengths to advance the science of remote sensing, especially as it applied to first responders and environmental solutions. They have also worked toward developing many data-capture and analysis concepts used by departments of transportation, departments of natural resources, and search techniques used by first responders nationally.

Ray,

What a journey we have had, culminating in developing a way to bring back the beast. Over the last ten years you and I have had the most enjoyable stories, dreams, and nights of chaos, adventure, and often downright humor. When your life is a story, search for the amazing in it. You have taken a wild ride, finding joy in the adventure. We started off with an overnight car ride, looking for a gas station in the early morning on an empty tank. From there we developed an incredible idea, pursued it, and were in our own way successful. We developed methods no others thought about. We researched, got county governments in trouble, stood in front of hundreds of first responders – ultimately saving lives. We got a free private plane to an island, designed methods that are best practices in aviation safety. We about died in the sand dunes of Michigan, flying the mother of all drones. We were mistaken for multi-millionaires, brainiacs, and crazies. We even looked for Sasquatch, and designed a way we could actually once and for all discover his whereabouts. In the end, we were two guys just looking for adventure, meaning, and friendship. We found all of those, culminating with this book. I love you dear friend, it's been a ride for the ages. *Jeremy*

Ray and Jeremy

Ray filling up at a closed gas station at 3am (this trip started Ray and Jeremy's fascination with remote sensing)

Ray testing out his Near-Infrared (NIR) camera

Boarding a plane to Beaver Island Michigan to do drone research

Ray and Jeremy fixing the NIR camera at a wheat field in Flint, MI
(Ray ultimately crashed the drone this day)

Ray Searching for Sasquatch

Jeremy, Ray, and Eric Searching for Sasquatch

Ray and Jeremy find the problem utilizing thermals

Michigan Dunes in NIR

Ray flying his drone with NIR camera on Michigan dunes

Ray and Jeremy testing a drone in 26 deg below zero

Jeremy and Ray researching sugar beet processing best methods

Ray piloting

Ray and Jeremy help create a spectrum analyzer drone

Ray fixing his drone on the shore of Lake Michigan

Searching for Sasquatch

Discussed in Chapter 4 of this book and filmed on scene of a sasquatch sighting in Northern Michigan, this pilot episode takes viewers through the journey of looking for the elusive creature.

NOW STREAMING

HOW WILL YOU HELP US BRING BACK THE BEAST?

1. _____

2. _____

3. _____

4. _____

5. _____

For more information visit www.SavingSasquatch.org

Made in the USA
Columbia, SC
29 September 2024

43292716R00041